Secrets of the Squaw Bay Caves

by

Mike Savage

DISCARDED

Box 115, Superior, WI 54880 (715) 394-9513

Secrets of the Squaw Bay Caves

Second Edition

Copyright 2002, Michael P. Savage

ISBN: 1-886028-25-7

All rights reserved, including the right to reproduce this book or portions thereof, in any form, except for brief quotations embodied in articles and reviews, without written permission from the publisher. This is a work of fiction. Any resemblance to real persons is purely coincidental.

Published by:

Savage Press
P.O. Box 115
Superior, WI 54880

Phone: 715-394-9513

E-mail: mail@savpress.com

Web Site: www.savpress.com

Printed in Superior, Wisconsin

Learn to love the questions.

— Rilke

Five miles east of Cornucopia, Wisconsin by water, a brave outcropping of pure sandstone shrugs its majestic shoulders against the largest body of fresh water in the world. Being one of the oldest geologic formations on the planet, the massive cape has resisted the throes and thrashing of Lake Superior for eons. But there have been concessions.

The layered red rock of the sheer cliff faces, in addition to being relentlessly pounded by waves large and small, has been infiltrated by the pure water. Though pure relatively pure chemically, the water has, when freezing and thawing, what could be construed as a decidedly impure, perhaps even evil, effect. It has destroyed the cliff's integrity. For a couple of miles around the periphery, all along the water line, hundreds of caves of every shape and size have been naturally excavated into the porous rock. Each structure has been formed entirely by the pure artistry, or devilry, of Nature's hand. It is a place of wonder, mystery, and awe.

This acne of sandstone has produced caves so deep that, swimming to the rear, all light is lost. Some caves are big enough to pilot a cabin cruiser through. Some are so small; a swimmer's hand cannot enter. Some are below the water line. Many are connected in a labyrinth. Swimming into one opening might lead to a dead end or it could reveal a passage through the solid stone to another tunnel, allowing the swimmer to emerge in a completely different area of the maze. Spelunking in trunks.

Nobody knows this, but one of the swimmable caves leads deep into the bowels of the bedrock. Once a millennia a swimmer is drawn to that particular cave for a meeting. For, far below and far in, an old woman lives. Her name is Ione. She is ancient Ot Chip Wa, an Ojibwa, a Chippewa. She is Anishinaabe.

When Ione heard that white shamans with the strange sounding names of Allouez and Baraga were wandering the land, she immediately paddled her canoe from Siskiwit Bay

Caves

to the Waanzah's. She'd had a vision. Sitting on the sand beach mending a gill net. She had dreamed while awake. Her reverie showed that, when the Chomoke shaman came, Turtle Island and The People would be tested. A giant trout with the gleaming silver sides raised itself out of the shallow beach waters and said to Ione, "You will surrender your life above ground so that the earth will survive. As long as you remain below, the earth is safe."

How do I, a chomoke, know these things?

Three years ago our family made an outing to the caves. We crowded onto my nephew's Bayliner and, laughing in the sunlight, cruised to the caves. We picnicked on the huge flat rock beside the mammoth cave mouth known as Cathedral Cavern and shouted happy greetings to the crabby sea kayakers who happened by. It was clear they wanted the caves to themselves, didn't want powerboat trash defiling the purity of their cave experience. The water was flat and warm Three of us dove in and swam after the kayakers. Their leader scowled as we stroked up to them in the emerald water. He grunted the group away from the porpoises. We laughed and splashed each other and reveled in our affinity to the fish. I felt superior to mere paddlers, sorrow for their separation from the water.

Everyone thinks the black gaping mouths are just holes eroded into the cliff face. I always thought that too. But, as I swam through one of the smaller openings, a rogue wave appeared from nowhere. The surge grasped me like a large powerful hand and dragged me away from the light. Fighting against the attacker's superior strength, my arms and shoulders flailed into the rough sandstone, which ground my skin raw. Darker and darker it became. My lungs screamed. My ears pierced my skull with repeated thrusting knives of pain. The swift current carried me onward for what seemed like hours.

The sunlight vanished. I knew I was going to drown. My head struck the top of the cave. Brilliant flashes of light spattered across the blackness of my mind. The scattering light sounded like a fresh walleye filet dropped into the smoking-hot grease of a black cast iron skillet.

So this is what it's like to drown, I thought. One of our family's mantras is, "Drowning is the best way to go. It's just like falling asleep." I'd heard it so many times I felt suddenly, rather peaceful. However, I didn't expect the brilliant explosions of light. I didn't anticipate that the chest pain would subside. In the midst of the field of erupting geysers of light, a tiny campfire flame appeared at a great distance. "Go toward the light," my nephew Billy had always told me. *I am dead and going camping in heaven*, I thought.

The campfire flame grew larger. I felt a chill rush up from the bottom of my feet, across the backs of my legs, up the small of my back, between my shoulders, into my neck. Like an arrow penetrating the chest of a small doe munching apples under a tree, the chill pierced painfully into the base of my skull.

A deep rasping sound grated against my eardrums. The sound was like a March-spavined Rez horse heaving after a winter of eating moldy hay. Below the pain being inflicted into my brain stem by the cold psychic arrow, I realized what the rasping sound was.

My breathing.

I wasn't drowning.

A slap of water sloshed into my mouth, I gagged and was submerged again as if some powerful sea serpent was dragging me to its lair. My lungs reprised their scream of protest. A second canopy of brilliant light flashes swirled across my cranium like hyperactive Northern Lights. The display cavorted around a central yellow flame. Again, what I thought was to be my last thought on earth entered my mind: *Thank you God for life.*

The central firelight grew larger. The circling trails of peripheral light slowed and spread out like a blanket being fluffed across a bed. The blackness wavered, then steadied. As the darkness stabilized, a million pinpoints of flickering white appeared like stars in the night sky. Again the pain in my chest receded. I felt a distance from it. I ached but did not suffer. My attention was entirely with the flame. The blaze was not threatening. Its heat comforted.

Suddenly I could breathe again. Gasping, my vision grew clouded with a red haze. The malevolent wave that had grabbed me and transported me against my will released its grip.

Sand.

I felt soft sand below my hands and knees. Crawling forward, I tried to discern my surroundings. All was red.

"You're bleeding," a woman's voice said.

I put my hand to my head. Hot blood and a large flap of hairy skin flipped backwards. "I've been scalped."

Laughter unlike any I'd ever heard saturated my soul. The sound was many dainty bells chiming together in harmony.

Then a hand touched mine. Warmth flooded me and dismissed the Lake Superior cold that had penetrated to my guts. Intense energy swirled around my cut like a thousand honeybees dancing across the wound's blossom. The red veil thinned. I knew the bleeding was slowing. The tingling in my scalp subsided. Gradually my vision returned.

What I saw was an old woman's face. It was brown, wrinkled and framed with jet-black hair that hung in thick braided ropes on either side of the kindest face I had ever seen.

"I am Ione," her voice said. Was it a real voice? Had her lips moved?

What I did know was, I was alive and on the sandy shore of a subterranean cove feeling pure love fall into me from the most magical face imaginable. It was Niagara Falls filling up a thimble.

Ione

Her face was round. Her lips were curved like the gentle bend of a July garden snake slipping between immature corn stalks. Her deeply lined flesh looked soft. Her eyes were pupil-less and so brown they were almost black. She wore a deerskin kuspuk with intricate, multicolored beadwork around the neck and shoulders. Her right hand was wrinkled and twisted. Seeing that malformed limb, I felt as if a flame were searing my own flesh. I winced automatically. She held the gnarled hand up. She placed the other hand next to it. Holding them both before my eyes she said, "One hand reached into the flame, the other pulled it out."

In my mind I saw the garden snake. The orderliness of the growing rows of corn quieted my child's fears and inspired the small kernel of courage in my. I ventured in where the stalks were taller than I. It was the Amazon jungle. I was an intrepid adventurer. Then the stick on the ground turned

into a snake and slithered away. I yelled in terror. The other kids came running. Bug Head caught the snake. He held the reptile up to his face and claimed, "This thing's been killin' our chickens!"

"Then *we* gotta kill *it!*" Little Prick said.

We took it to the stump of the willow tree. Nutless held the tail. Germ Face pinned the head with a stick. Stretched out across the stump, the snake's back gleamed. Bug Head beat the shiny back with an old gray board. After many blows the body separated.

Ione's statement about the behavior of her hands triggered that memory and exposed a wound that had never been salved. The snake had startled me. My outcry caused it to be tortured and killed. It was wrenching to see the beautiful skin destroyed. The blue and green was brilliant against the butter yellow belly. There was an iridescent purple hue that glowed from the combination of colors. The segmented pattern of its body was inlaid in patterns more intricate than a thousand-piece jigsaw puzzle. The orange iris of its eye was intelligent looking. The animal's integrity and wholesomeness seemed so important after it had been bludgeoned into gore. Ione's words uncovered my guilt over Snake's death.

"We each are the two hands of God," Ione said and she spread her own hands from before my face, swung them up to the wound on the top of my head and touched me. First she used the damaged one, then the unwithered. Then she kissed the wound. Another surge of white-light energy Niagarad from the kiss, through my skull, face, neck and torso. At my heart the energy drove itself into a hidden chamber. Long ago, at the willow stump, I'd filled that chamber with shame, locked the door and thrown away the key.

At that kiss, Ione's energy burst the shame vault of my secret. The door of the vault exploded at the touch from the stream of white light. A spray of putrid black gore containing dismembered parts of snake erupted from the black room. The white Ione-energy and the stinking blackness boiled and sizzled in battle. The goodness kept streaming down upon the blackness. The stench slowed to an ooze.

I heard the sound of distant crying. I looked to see if Ione was weeping. She looked at me with serenity.

The weeper was I!

"Snake forgives you," she said.

"I wanted to save it," I blubbered.

"It knows," Ione said.

My sobbing was uncontrollable, yet I had to speak. "I, I, Ungh. Ungh..."

"Take a deep breath," Ione said as she wiped my tears with the deerskin. It was soft as a whisper. She touched my wetness. I saw that her withered hand had become smooth looking. She pulled my head to her bosom. I felt the deepness of her breathing. I imitated her. The tightness in my own chest diminished.

"I was afraid of Snake and the other kids."

"You were just a little boy. You were powerless."

When I heard her words I saw my heart room again. It was changing from a strong dark cellar of stone covered with moss to a pleasant log cabin in a peaceful forest glade. Bees and butterflies flew gracefully from flower to flower. Soft yellow sunlight, filtered by the many majestic white pines gathered in royal array around the cabin, descended lightly.

The wracking sobs in my chest weakened. The headache at the top of my head had fallen through the layers of brain and dissipated like a soft spring rain. "I...I'm hungry." I sounded like a 3-year-old.

To my amazement, Ione didn't sound angry when she answered, "Let's have a little something to eat. Shall we?" She lifted my head from her embrace and looked me in the face.

I've never seen such a gaze. Something flowed from her face. It was as if a river moved between us. I wanted to hide.

Ione smiled. "You don't have to hide," she said. "I love you. All is forgiven. Let's eat." She moved toward the campfire.

I sat awash in new feelings. I felt clean, safe, loved. I felt as if there was room for me in the cave. I felt as if I could save the snake if I had another chance.

The Fishes and the Loaves

The transforming wash of love and acceptance surged through every cell of my body. The sensation was a wave that coursed up and down my torso; it surged into my limbs and eddied around my organs. My senses followed the healing tides as they moved in their mysterious patterns.

Then my perception changed. Instead of perceiving the grand sweeping movements of the healing, my inner vision started a type of descent. It wasn't falling face forward. It wasn't even going downward, really. It was a receding, a racing away from the larger theater of action. I flashed past flesh and bone, connective tissue and body serum. My vision blurred as if there were tears in my eyes. Deeper and deeper I went. Smaller and smaller I became. Past arteries and blood veins and beyond corpuscles. Then I was above a single cell.

It trembled in the velvet blackness that was the deepest interior of my body. The cell was anemic and pale white. It luffed like a main sail on a virtually windless day. A faint dot occupied the center of the cell.

To the left, motion. I turned. A great dark wave was rolling in. The sheet of my inner darkness was rising up and rolling toward the cell as if a tall rolling breaker was coming in from the middle of Lake Superior to wash up on the sandy shore of Siskiwit Bay.

It was the healing surge from Ione.

The wave was alive!

Its lifelike quality transfixed me as it washed across the cell. I followed its mesmerizing motion to the right as it vanished into the oblivion of me.

After the wave was gone for a long time, it occurred to me to look down again. A wholly different vision greeted me. The cell was changed! A vibrant rose color shone from it. There was a dark, healthy looking nucleus now. The cell wall was thick, showing contours of definition akin to a weightlifter's sculpted muscles. The wall looked like a barrier of solid stone that had been constructed by ancient and powerful Celts who had manhandled each stone into place by the shear brute force of their own mighty bodies. And the energy! The whole cell hummed with verve. It throbbed. It pulsed. It vibrated. It resonated with a ringing of such clarity; I was reminded of the bell-like quality of Ione's initial laugh.

At the thought of Ione, I left the cell. The departure wasn't directional. It was just movement. The journey back was not the way I had come. I whooshed past a broad field with green, ankle-tall grass. In the background, a huge stone castle loomed up through the morning mist. Around the perimeter of the field, tents, each with a coat-of-arms flying in a mellow breeze. A gallery of observers was watching two knights fighting.

Next I passed a farm family in a homestead's orchard. Mother, father, daughter and younger brother were harvest-

Caves

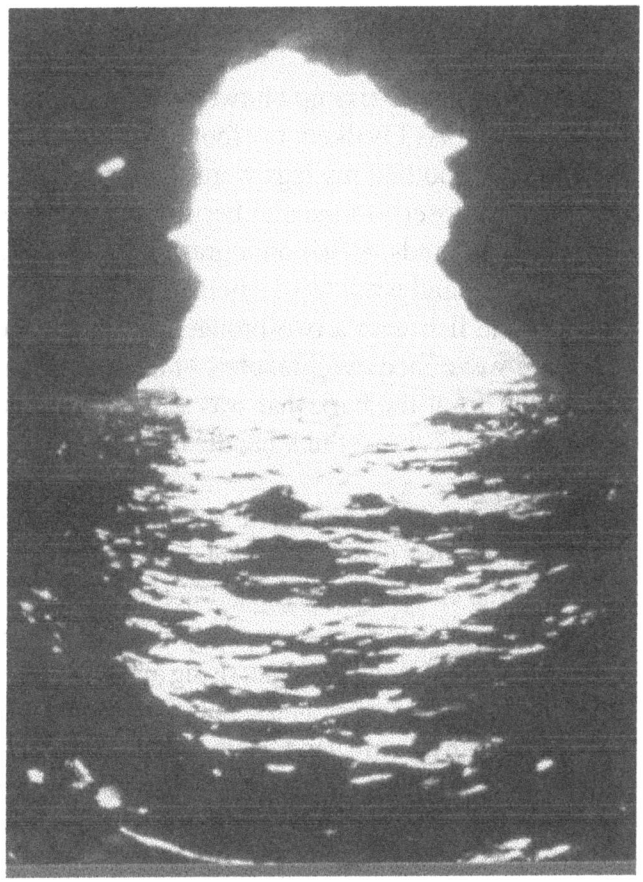

ing apples on a crisp fall day. They were all smiling and taking turns talking about the dreams each had had.

Next I saw a large stone oven in a meadow. A plump woman in a dirty cotton apron was pulling a paddle of scones from the open hearth.

The baker receded from sight and I heard humming. A swarm of bees flew past. Their humming pleased me.

Tableau after tableau wheeled past in a grand panorama of wholesomeness until I came back to the cave. Ione. As I studied her, she turned. Her face was all wrinkled now. Her lips were thin and her smile weak. Her skin looked like paper folded into a fan of a million folds. "Come. Eat," her voice croaked.

I stood. It felt good to rise up on two legs. The sand was soft against my feet. As I walked, my footing was unsure. As I walked I realized too that my legs were weak at the knees. Weaving my way closer to Ione and her fire, I saw beneath the mystery of her hands, a fish on a grate. Both her hands were now shriveled and wrinkled. I smelled the fish cooking. I saw her turn the fish with a two-pronged, carved wooden fork. The tines were once two branches of a maple sapling. This I could tell from the bark that was still on the crotch. The tool's shank sank into a thick handle of pure silver with intricate etchings.

I fell to my knees in the sand. After Ione turned the fish, she placed the primitive fork in the crack between two of the larger rocks of the fire ring. She raised her withered palms above the fish and closed here eyes. She pursed her lips and straightened her back. Her chest began rising and falling rhythmically. I too closed my eyes. The second they were shut; I heard Ione's voice inside my head. With my ears I could hear the faint little kitten-kisses of the water lapping on the shoreline behind us. But in my head I heard Ione's voice instructing me.

What Ione Said

This first fish is Acceptance. When you eat it you are ingesting the most important meal. The fish accepts its role. If it swims and dies of old age, it is accepting. If it swims and is chosen to feed a visitor in a cave, so be it. This fish also gives the knowledge of limits to all who eat it. The flesh of this fish instills the essence of fishness in the eater.

I looked at Ione. Was she insane? What was she talking about?

The old woman laughed. This laughter wasn't the bell-chime laughter. It was a first-grade-teacher-amusement-with-an-innocent-pupil laugh. Then her lips moved and I heard her voice in my ears. "The Iroquois ate the heart of their victims to acquire their power."

I looked at her, a deer in the headlights.

Ione sighed. "The fish lives within its fishness. It does not try to be an otter. It knows its limits. If you eat this fish you will acquire the knowledge of the fish, the awareness of your limited abilities. You will accept that you are not God. You will learn to judge only yourself and not others."

"I've never thought I was God."

"Everyone thinks they're God."

I wanted out. At first, being alive and not drowned in the caves seemed like a good deal. With this crazy woman talking in riddles, I was ready to ditch.

Ione sighed again. "The fish will teach you when you eat it."

She retrieved the fork and plucked the fish from the grate. She slid it onto a gray kiln-fired stoneware plate and handed the offering to me. "Take, eat, this is the body, given for you."

I took the plate and looked from the fish to Ione and back to the fish again. Suddenly I was famished. I put the plate on my legs, which were folded under me on the sand. I grabbed the trout and tore the skin off. I stripped the warm flesh off the bones and greedily devoured all the flesh I could pluck with my fingers.

All the time I ate, my attention was wholly on the food. Never had I eaten such a succulent meal. When, with trembling fingers I seized the last morsel of flesh and put it to my lips, my attention returned to Ione.

She was young now! Gone her wrinkles! Her face was smooth, her lips full. She smiled and stood. Without speaking she walked away from the fire to what I guessed was the south. She walked slowly, but the darkness enveloped her quickly. An eviscerating fear arose in me. She was leaving! I tried to get up quickly. Shards of pain like glass breaking shattered through my knees. It felt as if I had been kneeling in prayer for days. Ione's back vanished into the darkness. The instant she vanished, the fire went out. Consummate blackness virtually slapped me in the face. It was cat-sleeping-on-the-baby's-face blackness. Fear seized me like a giant bear squeezing until my guts squirted out.

Despite the agony in my knees, I commanded my legs to work. I hobbled in the direction I thought Ione had gone. Stumbling forward my feet found greater and greater purchase. The sand became solid stone and, plunging forward; I heard the slap of my bare feet against smooth stone. A chill overcame my skin. My entire body trembled in the cold. The frigid-force reached my body center as I blinked my eyes against the darkness. I tried to see some light. It was so dark, the effort to see hurt. Plastic-bag-and-overdosed-in-the-tub blackness grappled with my soul. Complete disorientation swarmed around me like a million biting piranhas. Which way was Ione? Where was the water? Was the cold stone cave wall ten feet away or an inch from my nose? Then I imagined my nose broken and bleeding from smashing into the rock. I stopped dead.

The temptation to lower myself to the cave floor and curl up in a tight little ball crossed my mind. If I hadn't just eaten, I would have given up, laid down and died. But there was energy to continue. Why not move a bit? Why not raise my hands, search the inky blackness in front of me? I weaved my hands outward like a Preying Mantis walking along an alder branch. I took a half step forward.

I proceeded like this for... how far? How long? Was I moving in a circle? There was absolutely nothing to aid me in discerning reality. The initial anxiety of my abandonment subsided as the routine of movement distracted me. The exploratory movement imparted a sense of progress. There was absolutely no sensory evidence to support the feeling that I was advancing my cause. Fro what seemed hours or weeks, I simply had to accept the fact that it felt like I was going in the right direction, that I was acting in my own best self interest.

After much cautious groping, my right foot struck something soft. I sank slowly to my knees in the darkness. Something warm. Fumbling with the object told me I was touching soft leather with beadwork and sleeves. It was a garment of deerskin. In my mind's eye I imagined it to have the same beauty as Ione's jacket. As I groped the clothing, I found two additional pieces. These I explored until it was clear they were moccasins. The supple texture of the garment and footwear reassured me. I concluded that the garment was a one-piece smock with long sleeves. I struggled to don the garment. Smoothing the supple leather across my body, I found a tie in the middle, which I drew tight.

Instantly warmth rushed across my skin. My feet grew cold in comparison and I rushed to boot up. As I tied the footwear's leather bindings, it seemed as if I was actually seeing the fine beadwork, not simply imagining it. Was there some sort of faint yellow light showering down on my new footwear? I raised my head. Utter blackness. I lowered my eyes. There *was* some sort of glow. In that darkest of dark, I held out my hand and for the first time in my life, perceived my own aura. I radiated a golden luminance. As I sat on the cold rock and laced up the ankle ties of the half-boot moccasins, confidence grew in me. I would be able to light my own way now.

After tying the last lace around my ankle, I looked up and tried to see. My light was so feeble. I'd never be able to light my own way. Panic reignited my belly. I couldn't even see a cave wall or a pathway or anything. With my arms still splayed out ahead, I brushed against the rough sandstone walls of the passageway. A gruesome mental image of me clawing weakly at the hard sandstone with bloody fingers trying to escape, dying from exposure and starvation, flashed into my head. It was the last desperate thought of a terrified man.

Then I encountered a corner, negotiated it and a great relief flooded my soul. I saw a faint blue luminance ahead. Ione! She was moving, leading me onward.

I followed for what seemed like hours. Or was it minutes? Could it have been days? Weeks? All I knew was Ione's azure radiance in the distance. All else was blackness. The floor—didn't exist. However, by the ache in my thighs I had to conclude that we'd traveled far... many miles. Despite my disorientation, it felt as if we were beneath Grandview, Wisconsin, thirty or forty miles from Lake Superior.

The statue of Chief Nokomis in that town's park came to mind. History has said that Nokomis paid his debts with silver he dug from a secret mine. The silver handle on Ione's cooking fork came to mind.

The next think I knew was, Ione and I were in a solid stone room in which another fire burned without smoke and had no apparent source of fuel. On the grate was another fish.

The Second Fish

Again the unsettling experience of hearing Ione's voice without the benefit of ears. She was communicating directly with my spirit. The second fish is Respect. If you show respect to everything, all will be safety in your life. There is no danger in the Universe that can overpower respect. If a bee flies up, you pause in fear of its sting. This stillness is respect for the bee. It will not sting you.

My lips moved automatically, "What if it's a woman with a gun?"

Ione gazed into the flames. I expected her to look at me, be distracted by the noise of my words. She was statue still. Again in my head the disembodied voice. "If you'd shown yourself enough respect in the first place, you wouldn't be confronted by a woman with a gun."

Derision rose in me. I thought, *what do you know about modern life? You've been hiding out down here for centuries.*

Ione's immobility shifted. It wasn't really movement. It was more of a waver in her energy field, but because she had been so till, discerning the displacement was easier. This time when she spoke, her lips moved. "The further away I get from modern life the more I become acquainted with it. I came here hundreds of years ago when Father Baraga arrived. I paddled the canoe across the big water and arrived at the waanzhs wet and weary. The speaking fish appeared again and opened its mouth. I got in. The fish swallowed me and swam forward through the ages. I saw everything you call modern. Then the fish brought me to the caves. The fire was already burning.

"Since that time I have seen the outside world in visions and dreams. It is the same as when the fish shared its vision travels. The technological dilemma the world faces requires a new meal of fish and bread. That is why you called to me."

"I called to you?"

"Yes. Deep in your core, in the place where you know you are loved and know that you are love, you want to heal the world. When we reach your journey's end, you will bring back all you have eaten here. As you live out your days, you will feed this food to those you meet. The cure will be small and seem powerless. It must be this way in order to humanely contaminate those who live constantly with the poison of self-hate. Beside a large dose would be detected and countered. As each person you feed grows, the healing dose will enlarge and change all."

Ione then looked into the fire. I followed her gaze. In the flames a vision appeared. A young Indian boy, long brown hair hanging to the middle of his back in a ponytail, stood on a bulldozer. He was pouring a ten-pound sack of sugar into the fuel tank. Finishing his sabotage, he jumped to the ground. He stuffed the empty sack inside his jacket and said, "Let's get outta here."

The vision scene enlarged. Instead of seeing the vision in the flame, I was in the vision, standing next to the young saboteur. "Fine by me," I said proffering a fish cough drop. He eyed the lozenge, fidgeted a couple of seconds and took it. The renegade unwrapped the medicant. I looked around. Above us a vivid canopy of stars framed the blazing Milky Way. In the starlight I saw we were on a typical logging landing in the middle of a Jack pine plantation. As he put the drop in his mouth, I smiled. I'd contaminated yet another with a drop of respect. His life would never be the same. Then a vision within a vision occurred. The lad was before the Bayfield County judge agreeing to make restitution to the logger.

Then I was back in the cave.

Confused, I looked at Ione. "Me?"

She simply smiled and said, "Sleep."

The last thing I remembered was trying to figure out if my ears had heard her command, or if it was mental telepathy again.

The Third Fish

When I awoke there was a small solid silver chalice half filled with water sitting next to the fire ring and grate. Ione wiggled her fingers in the water of the chalice. A fish came to the surface. Ione lifted the trout from the water and slid it into my hand. "This is Forgiveness," she said, adding, "There must be conflict between all people. Wars and rumors of wars are as essential as eating, for, without pain and conflict, forgiveness is impossible. This fish's purpose is to move each eater through the cycle of grace. Grace is the blossom of forgiveness which is the result of respecting all and accepting."

I looked at the two-pound trout in my hands. Its mouth

opened and closed. Its gills swelled open and shut. "It's so beautiful."

"Kill it," Ione said.

"I can't."

"It must die so you can live."

Despite my conviction against killing such a gorgeous creature, Ione the eternal shaman woman was my only hope of getting out of this labyrinth alive. I raised the fish to bash its head on the stone floor.

A split second before I acted, I realized I wasn't hungry. I saw inside myself and found no hunger anywhere. I looked in my stomach and in my soul. Nowhere within did hunger exist in any form.

"I can't do this. It's disrespectful," I said.

"The fish will forgive you. It is Forgiveness. Kill it," Ione said.

Instead I reached over and slid Forgiveness back into the goblet.

"You have done well. I forgive you for disobeying. You accepted a greater command, the dictate of your own inner truth," she said picking up the goblet. "You must handle Forgiveness again," she said handing me the container. Intricate etching of three fishermen, a net, a rising sun and birds overhead were traced around the outside of the high-gloss silver exterior.

I looked into the goblet. Forgiveness floated belly up. It had been out of the water too long.

"Breathe life back into in," Ione instructed.

CPR on a fish had never occurred to me. Never the less, I scooped Forgiveness from its coffin. Holding it up to my face I looked at its aquiline features. The thought of kissing the fist was repellant. I did what came to mind. I blew it a kiss.

Ione laughed.

In my hands the fish came to life. I felt a wave of life tremble across its body and transfer into my hands; travel up my arms and back into my chest. The fish blinked. It seemed to smile.

"The breath of life," Ione chuckled.

I put the fish into the container again and its tail vanished. I peered into the depths of the goblet. The fish was gone. I stuck my hand in. Nothing but water.

"Drink," Ione said and she stood and walked back the way we had come.

I wanted to follow her, but my mind was suddenly fogged in. I knew I had to follow her, but something wasn't right. There was some sort of prohibition against following her. She grew more distant. It was like looking through the wrong end of binoculars. I knew that when she vanished the fire would go out. What was going on? Why couldn't I follow her? She grew so small. She was just an inch high. The smaller she got, the more of my surroundings I took in. The chalice! It was so heavy. It had to be lightened. Raising it to my lips, the chalice grew heavier. Struggling to tip the contents into my mouth, I looked into the cup and saw millions of fishes swimming. The second the fluid touched my lips I knew it wasn't water. It was wine. The instant I put the silver stemmed goblet down, the brain fog cleared and my vision returned to normal. Ione was just going around a corner exiting the room. I bolted toward her. I did not want to be in the dark again.

We walked forever. Or was it just a few seconds? An hour? A day? Ione's aura dispelled the darkness, but it was such a small bubble of light. It was such a capsule of isolation. The isolated feeling changed when we turned right. Another long period of silence ensued. The only sound was the scuff of our steps against the stone floor. My eyes had either grown accustomed to the dimness or there was more light present, for, I found that I could see walls. Slowly the realization dawned on my that there were wall drawings. They emerged from the gloom to appear on the right and left. We never stopped to appreciate the illustrations.

This failure to stop and observe caused me to catch one of the illustrations in a sort of snapshot type of view as we passed. I only had a few seconds to view the image, yet the sight remains clear in my mind's eye to this day. A comet was crashing into northern Wisconsin. Lake Superior was being splashed into space like it was a small mud puddle being stepped on by a Clydesdale. From the cataclysm, four horses with four white woman riders were heading in opposite directions away from the impact. Alongside each of the mythical travelers horses, an Indian man ran. Each red runner held onto their horse's mane and kept pace with the galloping steeds by bounding along with great strides while supporting themselves with their arms. The technique of skipping across many yards of ground as the horses ran, touching down intermittently like a stone skipping across the water, looked masterful. How the artisan was able to portray such motion in a single panel was mystifying. Was the artist Ione? I wanted to ask.

As we walked, a sound was coming to my ears. I forgot my question. It was faint and distant, but I could hear a noise as if a temperate breeze was rustling through the dry leaves of autumn. I wondered if it was fall above ground. I wondered if my family had held funeral services for me. I'd been gone so long. No doubt they had given up on recovering my body. How long does it take for someone to be declared dead? I felt sad.

Then Ione's voice sounded between my ears, *Forgive yourself.*

The noise ahead grew louder. "What?" I asked. Had I heard her voice or was it more of her telepathy?

Ione turned a corner. I followed, awaiting her reply. What I got instead was a shock. There in front of us was a beautiful copper colored waterfall. I thought of Copper Falls State Park in Mellen, Wisconsin. At out feet was a dark pool of water. The pool was decorated with about thirty floating tufts of

ivory colored foam the size of volleyballs. The armada of meringue crafts circled around the surface of the stygian pool bathed in supernatural light. It was a beautiful sight. "You must forgive yourself for trying to protect everyone, for taking care of everyone's feelings by your own," Ione said.

"Why is that a sin?" I asked as I watched the ivory ships spinning like tops.

"Who said anything about sin?" She squatted and reached for a fluff of foam. She scooped a handful up. The rest of the flotilla circled nonchalantly, propelled by the currents from the splashing falls. I thought of the Kettle Hole on the Siskiwit river. Ione put the foam to her lips. "Yum. You should try this."

I stooped and scooped a foam ship for myself. It tasted like chocolate. As we squatted there a fish swam to the edge of the pool. It stuck its head out of the water and said, "Hello."

"I am Understanding. Me in your belly unravels mysteries. To acquire understanding, I must be eaten entirely. No filleting. Understanding must be all consuming," the fish said. It smiled and schlooped back under the water quickly.

Ione looked at me and smiled also. I fell in love with her. Without words she said, "Accepting the viscera as well as the flesh of this fish into our inmost being will bring knowledge to the cellular level. To partially ingest this fish and accept only the best flesh is to dishonor the all-ness of creation. Eaten entirely, this fish will reveal that God is not finished yet and that all mankind will eventually see the just-ness of allowing evil.

I wondered if I would soon be eating fish guts.

The ugliness of that prospect vanished when Ione stood up. I remained on my haunches. She reached down and grasped the edges of her smock. She grasped the garment and started raising the soft leather. I thought of averting my eyes. Was she too holy to see naked? I heard the garment floof to the stone floor. The shoosh of air from its landing wafted across my legs.

"You must look."

I was happy. I wanted to see her beautiful body, the curve of her form, and the shape of her breasts. I looked up and was terrified.

Ugly scars completely covered the flesh of her belly. I recoiled in disgust, falling backwards from the obscene looking flesh in a backwards crab scuttle.

Ione lowered her fingers tenderly fingering the ridges and jagged edges of the offending marks. "Injustice is a perceived evil," she said. She held her hand out toward me.

I knew she wanted me to touch those hideous scars.

"Do not be afraid," she said. Her voice was tender. It called to me as if she were a mother wanting to cuddle her child.

I could not resist. I did not want to disappoint her. Also, as revolting as the mass of scars was, I was morbidly curious. I wanted to inspect them, wanted to know what caused such damage. I approached cautiously.

She waited with her hand outstretched. She took my fingers in her own and placed them on her belly. "The children of truth," she said.

Touching the scars I felt a strange movement, as if they were alive. They murmured with their own life force. I drew closer. Then I saw. "The garden snake," I gasped.

Searing pangs of sorrow raced out to my arms and down into my groin. I gasped at the agony and jerked my fingers away. I hugged myself.

"Snake is wondrous happy in me," Ione proclaimed. She pirouetted and shouted, "YEOW HAW!" And dove into the tannin colored water.

I started slapping my fingers against the top of my thigh. The pain in my guts subsided with each whack. I fixated on the task of slapping away the internal searing for, what? Minutes? Hours? Then it occurred to me...how long had Ione been underwater? She had not surfaced since leaping from the shore. As if in answer, her head broke the surface. She splashed me, scooping liquid from the pool in my direction. I say liquid because it did not feel like stream water. It felt like warm bath water. It felt like thick glycerin water for blowing gigantic bubbles. "Come in," Ione commanded.

Part of me wanted to dive in, get close to her, rub my belly against hers, take the snake scars from her belly onto mine. After all, I'd killed the snake. I deserved to be scarred, not Ione. Part of me was afraid of the water. It was strange water. It felt warm instead of what I expected, cold. It felt slimy instead of squeaky. Maybe it was some sort of acid. Maybe Ione could stand it because she was a goddess. Maybe it would dissolve me and the flesh would slip from my skeleton like an Eskimo Pie falling off its stick in July.

"You don't deserve scars," Ione announced. "See?" She said and arose from the water. She pointed to her belly. The scars were gone, washed away by her immersion?

I looked around, ashamed to pull off my buckskin garment.

Ione laughed and dove underwater again. I pulled my kuspuk off and laughed out loud myself. I still had my swimming trunks on.

I was about to dive in when Ione surfaced and held up her hand like a policeman stopping traffic. She pointed at my trunks and said, "Mankind must have clothes, but truth swims naked."

I did not want to strip. I wanted to dive in without showing myself. I could disrobe once underwater. But, there was an invisible net restraining me. Ione averted her eyes. I jerked my trunks off and dove in.

The dive made me think of a circus tiger jumping through a flaming hoop. The water tingled my skin from the tips of my fingers to the soles of my feet. The sensations weren't altogether pleasant. It was somewhat biting. It reminded me of the time a battery exploded in my face. It was during a heavy rainstorm. A guy was stalled in a parking lot. I happened by. He asked for a jump, even had cables. When he connected, the clamp sparked. The next thing I knew, my face was itching all over from the acid spray of the explosion. The guy had gotten the worst of it. He pawed at his face. I stood back from under the hood, tilted my head up and let the rain rinse my face. The feeling of being eaten away and being washed at the same time, I'll never forget. A similar dual sensation applied in particular to the wound on the top of my head, the one I'd gotten upon entering the caves. The scar felt like a thousand hungry, but friendly, ants were eating it.

I swam around under the water for a week. Or was it a few minutes. Regardless, I was breathing under water. Then I surfaced and saw that Ione was ashore, completely clothed. I treaded water and watched as she held her hands in an inverted cup six inches above the rock floor. Beneath her hands a fire appeared. She raised her hands a few inches and a cooking grate and silver soup kettle appeared.

Then all around me the water became alive with fish. There were a million of them splashing and thrashing the water to a froth. She walked to the water's edge and smiled. She pointed. It was an impish point, with just a little dip of her finger and a slight bend of her wrist as if she were ploinking a single key on a silent grand piano. One fish leaped from the water in a graceful arc. It fell into her hands and the water grew silent.

Beneath the water's surface, all around me, I felt the rush of a million fish bodies. Feeling their life brushing against my skin was alien. I'd never been so caressed. So touched. A vision from an old Tarzan movie sprang to mind. The hapless safari was lost in deepest, darkest Africa. Starved and dying of thirst, the come upon a peaceful lagoon. And excited porter runs into the water. It boils with a million starving piranhas. Then all is quiet. A second later a ghastly white skeleton rises to the surface. Feeling the millions of fish sluicing across my skin, I saw my own skeleton bobbin to the surface of this suddenly hostile feeling glycerin pool. I tried to swim ashore. My arms swept through the water, but the fishes held me. The harder I strained against the offending liquid, the more tightly it grasped.

"Ughh," I cried.

Ione glanced at me from her work. "What have you done?" She asked calmly.

I could not answer. I was intent on extricating myself from the rapidly solidifying mass. In addition, instead of fleeing my thrashing, the fish were getting closer and closer. The

water! The fish! Me! We were congealing into a solid!

Ione didn't move. My struggles grew more frantic. I wanted her help. Why wasn't she rushing to my rescue? I fought harder. The water compacted and so did my heart. All this time she acted like my friend. Now the truth was clear. She cared only for herself. Then in an instant all around me was solid. The sensation reminded me of Finnegan's hayloft. I was playing *Combat* with Mike Finnegan and his younger brother Mark. We were in the hayloft of the big red barn. Mark and I had laid a trap for his older brother. We'd pulled up bales to make a hole and waited for him to come up from the cow stalls below. Then we pushed him in the hole, threw bales on top of him and started jumping on top of the bales. It was funny to us. We laughed and ran across the hay to the other side of the loft. Mike, who was my age, but much stronger and tougher than I, crawled out from under the trap. Mark, knowing his brother's capacity for payback, wisely fled instantly. I didn't run. Mike pinched my neck so hard I had to march over to the hole. He pushed me in, threw bales on top and started jumping up and down.

It was an evil place.

I couldn't move. Pain shot throughout my body every time Mike jumped. It was hot. The bales squashed me and pricked me with sharp-ended hay stalks. I could barely breathe. For the second time in my young life, I experienced the unique confidence that I was going to die. Then the pain stopped. The bales rose. I could breathe. I saw light. Mike held out his hand and helped me crawl out of my coffin. I couldn't stand. Mike held me around the waist with his right arm. I looked at Mike Finnegan. His face was somber.

"Why did you do that to me?" I asked.

"So you would feel how I felt."

His answer surprised me. It had never occurred to me that big strong Mike Finnegan, the toughest, smartest, fastest, strongest could be hurt like me, by me. From that moment on I loved Mike Finnegan.

There in the water trap Mike's treatment of me seemed monumentally compassionate compared to Ione's inaction.

She just sat there looking at me. "What have you done?" She asked.

Not one part of my body could move. I tried to move my lips. They were stone, solid stone. I thought my answer, *Nothing!*

"Then why is the water solid?"

I don't know, I thought my answer, adding, *I was just swimming and the fishes came around me and everything started getting thicker, and now I'm trapped.* Under my words I was wondering two things. How could I be breathing if the copper water had turned to solid metal? Why hadn't Ione done what Mike Finnegan did?

"I can't help you," Ione said.

He words generated a great pang that shot through the center of my chest. It went into my throat and tried to shoot out the top of my head. I thought of the story of the Iroquois warrior who, with a hatchet, split open the skull of a British captain as they sat around a campfire. The body was still upright and convulsing as the Iroquois scooped out a handful of brains and ate them in front of the others around the fire.

The memory increased the pain in my head. The pain redirected itself and shot down through my limbs. The agony was trapped and wanted out. It was banging around inside me like a BB in an empty tin can. Wave after wave of reverberating pain coursed through my body.

ARRGH! I cried out inside my true self.

Ione stood up. She gazed at me and remained mute.

The pain echoed forth and back, eventually subsiding. Inside I cried, *WHY?*

"Why what?" Ione asked.

Why won't you help me?

Ione lowered her eyes. Her voice was soft. If the waterfall hadn't been silent, I wouldn't have heard her whisper, "You didn't ask."

Rage burst in me. My body's interior convulsed with surges of hatred. I vainly tried to twist my head and swivel my neck. All I wanted was to be free. If I could get out, I'd find the way home myself.

Then a new water sound came to my ears. Was the waterfall softening? Starting to flow? I looked at Ione. She was crying. Tears flowed down her cheeks.

I hated her tears, hated her sobbing. Disgust flowed up from below my chest. As it rose, I felt my breathing grow shallower. My chest grew tighter. The vituperative sensation rose through my throat and flowed across my face. As it moved, my ability to think clouded. A coppery haze shrouded my eyes. I tried to speak, my jaw clunked. There was so little air, so little strength. In the mental haze, I envisioned the pool frozen over with my bronzed head sticking out like a turtle taking a peak for predators. On my face, sculpted for all eternity, was an expression of pure abandonment. A slight wisp of smoke trailed faintly out of my open mouth. Ione was not present in the vision.

Ione cried more openly and loudly.

The copper fog thickened. My chest grew more immobilized. I wanted Mike Finnegan,

Ione wept.

There was no rescuer, no helper. The tightness in my chest was no longer tightness. It was as close to immobility as possible. I recalled that moment under the hay. Death! Death was spreading across my face. It was coming from the inside! The stench! Ugh! My instinct was to recoil, to wretch. Only I couldn't react physically. I was bronze, metal, unable to be alive.

However, there was strength in my fear, my revulsion at death. My chest managed to puff out the meekest breath upon which a single syllable rode, "Helfff."

Ione stepped forward. She daubed at her tears with open palms. She kneeled at the hard water's edge and leaned forward, placing her palms on the coppery surface. A wet looking stain spread outward from her hands. A puddle of water formed between her hands. She lowered her head and lapped water like a beagle. Then she started wiggling her fingers like drumming on a steering wheel at a stoplight. Little splashes of water spattered her skirt. The index snake on her left had lengthened.

It was the garden snake!

It grew out of her hand longer and longer. Then is separated and essed across the hard pond's surface making an unerring path directly toward my left eye. The snake swam right to my face. It stopped and bunched itself up. I knew what was coming, could do nothing to defend myself.

Snake's tongue flicked out. Then Snake shot forward and burst into my eye. Another shower of midnight star-points burst across my mind as the snake's head pierced my eyeball. Snake plunged forward with increasing speed. As each segment of Snake's slithering body jogged past my eyelid, I felt the animal's power oscillating through my eye socket and into the rest of my body. Its segmented skin fluttered against my eyelid like a long chain being pulled out of a pickup box, each link chattering and jolting.

Through my right eye I saw a lump in Snake's body approaching. Was it a just swallowed mouse? No! It was the scar where we had mutilated it in half. When the bump got to my eyelid, Snake couldn't fit through. It fought to get in. Pain radiated across my face. Sharp tentacles of agony undulated across my temple, past my ear and up across the top of my skull. The back half of Snake flailed like a detached high voltage line in a tornado. Its body slapped wildly across my face and head. I opened my eyes wide and Snake's wound entered me with a popping sound. It was like jamming a quarter into a dime slot on a parking meter. The pressure of the passage made the frozenness in my faced shatter. Snake increased its speed of shooshing into me. The last of its tail entered and I envisioned it swimming through my blackness and out of sight into my murky depths. The second its tail disappeared inside me, I could start to move. Behind me, the sound of waterfall started cascading. The stream flowed. I scrambled out of the water and donned my trunks and buckskin. I felt freezing cold and joined Ione at the fire. She was placing the fish in the pot.

"Would you have let me die?" It felt marvelous to move my lips.

"Everyone's path is their own," Ione said. She reached into a pocket and pulled out the handcrafted silver and wood fork. "What makes you think you would have died?"

"I couldn't breathe."

"Why?"

"You saw the goddamned water. It was suffocating me!"

"You were suffocating yourself."

I wanted to kick her.

"Why do you always strike out when you are faced with your self?"

All the anger whooshed out of me. I sat down heavily lowered my head. "I hate you."

Ione sighed deeply. Silence spread across the cavern like a big quilt on a king sized bed in a master bedroom. The soup kettle bubbled. The scene of the fish jumping out of the water into Ione's hand flashed into my mind. The sensation of its million companions against my flesh recreated itself across my skin. I shivered. The scene of the piranha-flayed skeleton followed.

"You are afraid of what wants to help you and you embrace what wants to harm you. Fish want to feed you, not feed on you," Ione said softly. Tears flowed out both of her eyes.

I noticed again how dark and brown they were. I leaned forward. I reached across the fire. I touched Ione's cheek. The wetness spread across my finger. I put the finger in my mouth. Her sadness instantly spread throughout my entire body.

"Why are you so sad?"

"I grieve because my healing energy was agony to you while your own anger and rage brought you no pain."

I remembered how intense the pain of her energy was as it bounced around inside me. I remembered how familiar and friendly my urge to kill her was. I remembered that I was to eat the entire fish, guts and all, if I was to understand.

The Fifth Fish

Ione spoke without looking up from the mojakka, "Fish number five is unity. It is a fish of five colors. Its diversity is its wholeness. To reject this fish is to deny the glory of opposites. The smelt must school and feed on herring fry. The trout must gobble the smelt. The loon must spear and eat them all. We must consume the loon like a mallard dinner and fry the trout for supper. Then we must return our care and concern for all creatures. We must resupply creation every day until we die and ourselves return to spirit. All of creation is connected. Two people who would become one must be detached in order that the circle be whole and not lopsided." She looked up and added, "It is time to go. You must return to your totem."

Somewhere in the middle of my chest a superstructure caved in and my whole being collapsed. It sunk into a huge hole in the earth. It was sucked into a black hole in the galaxy. It evaporated like a mist in the Gobi.

"I want to stay here... with you."

"You were selected specifically because you would return. You chose to come because you knew you wanted to bring these truths back." The words flowed out of her mouth like a stream of Karner Blue butterflies and I knew. It was true. She was right. Only, I didn't care about the truths. I wanted to either stay with her or bring her back and show her off.

We started walking. The sound of the waterfall receded. The path felt all downhill. The murals on the side walls stopped appearing. The supernatural light dimmed. The blackness deepened. Ione's aura appeared. My aura glowed feebly also. But it seems a bit brighter then when first I saw it.

I wondered how I would explain my absence. How would I get home from the caves? Surely the search for my body had long ago ended. Surely it was winter. Could we walk home across the ice? Would our light buckskin outfits keep us warm enough to reach land and get a ride home?

Two things interrupted my ponderings: first, our entry into a long room. Comfortable looking armchairs and a sofa with a deer hide draped over it were hyphenated between numerous tables. There were coffee, end, dining and library tables. Each was loaded with books. Above one antique library table against a far wall were shelves about four feet long. On these sat many gallon jars of foodstuffs. I recognized macaroni, rice and garbanzo beans. The second was Ione's voice, "We must stop."

An inner knowledge advised me to lie down, but not on the sofa. I lowered myself to the sandstone floor and laid flat on my back. My buttocks and shoulder blades felt the chill of the cold sandstone. I shivered. Ione sat next to me. She crossed her legs. Her skirt stretched taut between her bent knees. I attempted to look between her legs.

She laughed. More chittering chime-bells. She spread her legs even wider. The leather garment stretched tighter. She opened her palms and pum-pummed the dress like a drum. "Oummmm," reverberated out of the darkness between her legs. The sound went out to the darkness and returned. "Ummmm," Ione hummed. Slowly she bent her right hand around her knee and below the makeshift drum skin. She stuck her hand up between her legs and pulled out a loaf of bread. It was a round loaf with a crown of five inches. It was hot. Steam rose from the browned crust. Sprinkles of almond shards were baked into the crown.

The Bread Loaves

She placed the loaf on my chest. She raised my right hand and placed it on the loaf. Heat surged up my arm. "There is no such think as all fish. Bread is Life. With each slice of bread we eat, love is multiplied exponentially. It is the least we can do, eating bread."

Ione trummed her skirt/drum again. She withdrew another loaf and placed it on my tummy. "This is the loaf of Silence. Even the fish, if he kept his mouth shut would not be caught. Silence begets wisdom, speech repentance." These words Ione spoke as she placed my left hand on the second loaf. Then she pulled a knapsack from under the couch. It too was decorated with exquisite beadwork. The scene depicted a fish jumping out of a yellow-bead sun rising out of Lake Superior between Eagle and Sand Islands. This vision was seen from the opening of a waanzh. Seeing the artwork, I knew Ione had watched thousands of sunrises from the cave opening.

Placing the first loaf in the knapsack, she said, "I cannot go with you."

Grief, Kitchi Gammi sized grief, sprang to life in my chest.

She placed the second loaf in the rucksack and I knew it was not time for her returning. Then a fish appeared in her hands. She put it in the knapsack. "It is your mission in life to feed these fishes and loaves to everyone you meet." She said this while packing three more swimmers.

The last thing she did was bend over and kiss me on the lips. Then I felt the embrace of the coldhearted wave whose duty it was to return me to the sunlight and I burst out of the water gasping for air.

"Ha ha, great joke," my nephew said.

"We thought you'd found a way to another cave," my niece said as we all treaded water.

"How long was I gone?" I asked, my head swiveling around to get a grasp on this new reality.

"About a minute," my nephew replied. "Come on! Let's go have a beer!" He shouted and he swam toward the flat rock and the remainder of the picnic.

There was no knapsack, no buckskin garment. I treaded water for a second, reached to feel the top of my head. There was a scar where none had been before. I swam after my family, my tribe, my totem and wondered if we would be eating fish or bread or both.